# GOAT FARMING AS A BUSINESS: a farmer's manual to successful goat production and marketing

## Oliver Blake

# 1.0 INTRODUCTION

In Zimbabwe, there are more than 3.5 million goats, 98% of which are native varieties that are kept by smallholder farmers. The majority of them are housed in Tse-tse infected regions and the drier agro-ecological zones of Natural Ecological Regions IV and V. Natural Region IV has long dry periods and occasional droughts due to its low rainfall. All things considered, goat relevance rises with decreasing rainfall. Goats are resilient animals that need little care and can endure under challenging conditions.

Goats are raised in vast agricultural environments, mostly for their meat (chevon) and, to a lesser degree, for their milk. These goats' low production is partially caused by a number of issues, including high infant mortality and poor animal husbandry techniques. Goats also offer manure for gardens and agriculture areas, as well as commercially valuable skins. Goats are raised for their wool (mohair) in different regions of the globe.

The expanding number of people on the planet is driving up demand for more meals high in animal protein. The goat may be a valuable asset in satisfying these requirements. This urges farmers to value their goat businesses by moving from producing goats for subsistence to producing goats for sale. Compared to big stock, it is simpler to raise the population of small ruminants, such as sheep and goats. Goat manufacturing has little opportunity costs from an economic standpoint.

*"Between 9000 and 7000 B.C., the goat was most likely the first mammal to be domesticated. This lengthy history of goat-human interaction demonstrates the range of purposes goats may fulfill."*

This handbook has been prepared to help farmers who are interested in beginning a goat business on a commercial scale as well as those who require information on goat husbandry. Although not all-inclusive, the material blends the experiences of writers and farmers.

# 2.0 ENTREPRENUERSHIP

## Objectives

**By the end of the session farmers should be able to:**
- Demonstrate the entrepreneurial skills required to manage a profitable business.

## Introduction.

An entrepreneur is a person who consistently spots market possibilities for goods or services, then creates new goods and services to meet the needs that are found. An entrepreneur must first envision a profitable goat company before allocating resources to meet the predetermined objectives.

The mindset of "Imbuzi ziyazibonela" (the goats should take care of themselves) permeates most farmers. They are unaware of goats' true value, worth, and the additional contribution they may provide to their lives if properly handled like a company.

The following are necessary for smallholder goat producers to run profitable commercial goat businesses: Realize that there are risks associated with starting a business; Have access to sufficient knowledge and information; Recognize opportunities; Make time and resource commitments; and Be ambitious but realistic in your goals.

## Qualities of becoming an entrepreneur

To become a profitable commercial goat farmer, one must acquire the following entrepreneurial skills: seeing business opportunities, taking calculated risks, setting goals, gathering information, adhering to a business plan, networking and persuasion, and methodical planning and monitoring.

## 1. Recognizing a potential for business

The secret is to be able to see opportunities in one's personal or professional life when others cannot. Hence, an opportunity is chance, an opening, or a potential that presents itself.

Therefore, anybody who finds resources, unmet needs, and issues in society and converts them into company concepts is an entrepreneur. Thus, the first stage in starting a company is coming up with concepts. A couple of farmers who see that the scarcity of beef presents a chance to sell more slaughter goats to each other, or a farmer who recognizes the necessity for supplemental feeding and proceeds to stockpile the feed and sell it to other farmers during the dry season, are examples of entrepreneurs. • A trader who purchases goats from small-holder farmers and sells them to abattoirs in Bulawayo and Harare; they band together, rent a truck, and transport fifty goats every month for sale in Bulawayo.

What goat-related businesses are there in your area? To find your business chances in the goat sub-sector, use the tool below.

| Problems | Business idea |
|---|---|
| e.g. Shortage of meat | Buy and sell goats to abattoirs |
| | |
| | |
| **Resources** | **Business idea** |
| e.g. Goats | Improve condition and sell to retail shops in Bulawayo |
| | |
| | |

| Unmet needs | Business idea |
|---|---|
|  |  |
|  |  |
|  |  |

## 2. Calculated risk taking

Taking a measured risk is required once an opportunity has been found and matched with one's skills. Finding a balance between success and failure is crucial when taking a risk. By gathering knowledge and making wise judgments, risk may be reduced.

The majority of goat farmers are afraid to take chances for the following reasons: they may lose their funds; they don't know whether the goat business will be profitable; and they don't know what prospects are out there.

Goat dealers are a great example of accepting risks. They do duties that the majority of smallholder goat farmers would prefer not to perform themselves. The expenditures incurred by the merchants include purchasing, shipping, pre-financing, payroll expenses (for buyers and herders), and fees associated with slaughtering.

Exercise: In your life, what measured business risk have you taken? Note the event down below. What took place and how did it occur?

……………………………………………………………………………

……………………………………………………………………………

……………………………………………………………………………

……………………………………………………………………………

..................................................................

## 3. Objective setting

An aim is characterized as a quantifiable and specified goal that must be accomplished within a certain timeframe and budget. Th[e] cornerstone of achieving goals is a well stated aim statement. Goals are intended to provide guidance, inspire hard effort, help wi[th] organization, and serve as monitoring instruments.

The following is stated in an objective:

- What goals are to be met?
- By whom?
- By what date?
- In what location?

Goals need to be SMART—specific, measurable, reasonable, attainable, and time-bound.)

| Example of an objective that is not SMART | Example of a SMART objective |
| --- | --- |
| "I want to have more money". | "I will sell 5 goats directly to TITI restaurant by September, leading to an increase in my income by 100 %. |

| Exercise: |
|---|
| Write a SMART objective for your goat enterprise: |
| |
| .................................................................................. |
| |
| .................................................................................. |
| |
| .................................................................................. |
| Taking your objective as it is, if achieved what would it bring to you? |
| |
| .................................................................................. |
| |
| .................................................................................. |

## 4. Information seeking

A company owner should learn about new technology and other aspects that impact their goat enterprise.

The data acquired will:

• Assist in lowering risk; • Help construct the company strategy; and • Give the entrepreneur the ability to make more intelligent judgments.

There are many approaches that may be used to get information. These consist of observation, questionnaires, interviews, and desk research.

Books, government documents, business magazines, radio, television, Internet services, business development service (BDS) providers like One-Up and Women-In-Business, competitions (fairs and shows), buyers, suppliers, other farmers, and newspapers are some of the information sources.

Exercise: How much do goats cost at the neighborhood sale locations?

What is the cost of goats in the closest city?

What choice would you make in terms of business with this information?
..................................................................................................
..................................................................................................
..................................................................................................

Farmer Magazines, books, pertinent government agencies, radio and television, electronic media, contests (fairs and shows), suppliers, purchasers, fellow farmers, and newspapers are some of the information sources.

## 5. Adherence to the company strategy

Once a business strategy is in place, it is necessary to:
- Maintain concentration;
- Be dedicated to the tasks at hand;
- Comply with a predetermined work schedule; and
- Encourage and lead the team members working on the goat company.

**Exercise:** Write your plans for your goat business for the coming 12 months.

..................................................................................
..................................................................................
..................................................................................

## 6. Persuasion skills and networking

You don't merely sell your goats to buyers or abattoirs. Reminding people that your goats or goat products are available is your duty, either as an individual or as a producers' group. You may do this by having in-person meetings with your customers, running advertisements, and staying in close contact with buyers and suppliers.

Strong commercial connections are built by networking and persuasion. Trust, interdependence, an equal power structure, a fair decision-making process, shared objectives, fair rewards, a problem-solving approach, and dedication are the foundations of strong business partnerships.

Exercise: Describe an instance in which you were able to convince customers to buy your goods?
..................................................................................
..................................................................................
..................................................................................
..................................................................................

List the different stakeholders you are interacting with and are relevant to your business.
..................................................................................
..................................................................................
..................................................................................

## 7. Persistence

There are several difficulties in the goat industry. In times of adversity, you must persevere. It is possible to convert a challenge into an opportunity. What's required is the capacity to endure and rapidly adjust to changes that occur. Additionally, one must be adaptable.

What did you do when some of your goats or children were lost?

.....................................................................................................

.....................................................................................................

.....................................................................................................

## 8. Independence and self confidence

When someone decides to start goat farming as a business, they must:

They should want to be financially independent, be in charge of their own destiny, believe in themselves, be able to make their own choices, and be self-sufficient.

## 9. Systematic planning and monitoring

Create a business plan that addresses the following in order to facilitate efficient planning and oversight of the goat company:

- Step1: Right now, where am I? (As of right now)
- Second Step: Where am I heading? (the intended state, the future, the mission, and the goals)
- Step III: What stands in my way of getting where I want to go? (The commercial setting)
- Step IV: How can I get there the most effectively? (Positive strategies)

- When will the scheduled tasks be completed, and how will I be able to gauge my progress? (Monitoring Schedule and Objectives)

# 3.0 Goat Breeds

**Objectives**

| At the end of this session farmers should be able to: |
|---|
| List the breeds that may be found in Zimbabwe and their characteristics. <br><br> Select the right breeds for their goat farming operations. |

## TYPES OF BREEDS

- The huge Matebele and the Small East African (SEA) goat make up the bulk of native goats in Zimbabwe; average birth weights for babies vary from 1.5 kg to 2.5 kg (up to 3 kg).
- The native breeds have excellent environmental adaptations.

Larger goats known as Matebele goats, with a mature weight of 45 kg, are found in the southern regions of Zimbabwe.

Other breeds that may be found in Zimbabwe include exotic varieties and the 65 kilogram adult Boer goat, which is mostly raised for meat. The Saanen goat is raised primarily for its milk, yielding an average of 3.5 liters daily. Additionally, Angora goats are used to produce mohair.

Boer goat                                             Saanen

# 4.0 Management of does and bucks

**Objectives**

> **At the end of this session farmers should be able to:**
> - Properly care for female and male goats
> - Know the age at first mating

In order to produce goats, both male and female goats must be given the proper care. This covers planned immunizations and dosages, additional nourishment, breeding stock selection, raising children, and weaning.

## 4.1 Management of females (does)

Young females need to mate as early as 12 months of age. An animal that is well-nourished will develop more quickly and be prepared for mating. Additionally, it increases litter size and fertility. Young animals that are mated before the age of eight months will be stunted for the remainder of their lives and will not reproduce well. A healthy mother may have children for around eight years.

Goats go through five months, or 145–150 days, of pregnancy. An adult female can only mate when she is ready, or when she is in "heat." The length of the heat phase is 24–26 hours. She should get the man throughout this time. Heat is generated in the flock by the male's presence. The animal's diet has an impact on when it becomes heat-ready. Symptoms that might point to an animal in heat include:
- Waving the tail;
- Hiding other creatures
- Looking for men

Symptoms Prolonged bleating
- Clear discharge

At least two months before to kidding, pregnant females should be kept apart from the main flock for careful observation. This lessens the loss of children as well. At this point, their bodies' nutrition stores will need to be enhanced with high-quality feed supplements. This will guarantee a child's health and enough milk.

Female goats (does) separated from the main flock

## 4.2 Management of males (bucks)

• It is known that male goats become pregnant sooner than female goats. Males in these situations need to be reared apart from females in order to prevent accidental mating.

- Bucks must always be fed and maintained in excellent health.
- Bucks with horns must be utilized for breeding in order to prevent hemoprodism (incukubili/bisexuality), which results from using hornless or polled bucks.
- Bucks may be chosen from a young age. For further breeding, a male child born weighing 2.5 kg or more may be chosen. Picking larger and rapidly growing bucks is a good idea.
- To enhance the likelihood of twinning, choose bucks from twin births.
- Castration or culling should be applied to males that are unfit for reproducing.

## 5.0 Breeding

**Objectives**

At the end of the session farmers should be able to:
- Understand different breeding systems
- Understand different mating systems
- Formulate their own breeding calendars.

## 5.1 Breeding systems

When it comes to producing meat and milk, the breeding system is a crucial component of goat production. It has a major impact on flock production both now and in the future.

Crossbreeding is the process of marrying two distinct breeds in order to blend traits from each breed and use the "hybrid vigor." Put simply, this indicates that the child outperforms the parent. One technique utilized in the production of milk and meat is crossbreeding. If done incorrectly, it might be catastrophic and cause the current genetic pool to vanish.

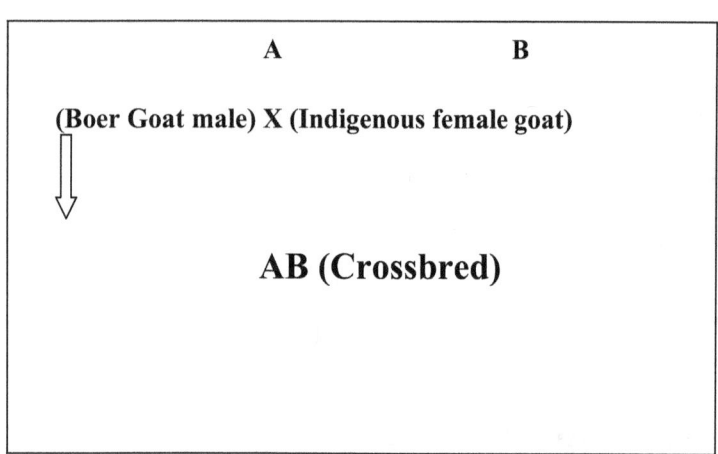

```
        A                      B
(Boer Goat male) X (Indigenous female goat)
    ⇓
              AB (Crossbred)
```

Pure breeding:
In order to preserve the desirable breed characteristics (color, size, meat, and milk quality), purebred females are paired with purebred males (bucks).

## 5.2 Mating systems

It is imperative that farmers possess knowledge of various mating techniques that they might use in their breeding flock. Allowing any number of bucks to run amok with a flock of ladies is known as random mating.

**Benefits of random mating:**
Advantages of random mating
1) Simple
2) Cheap
3) Since goats may give birth at any moment, a farmer can sell at any time.

**Disadvantages**
1) High inbreeding risk
2) A high chance of disease transmission.

The best females are paired with the greatest bucks via assortive mating. This is preferable than haphazard mating.
Advantages of Assortive mating .
1) Premium breeds
2) Preserve the genetic foundation

**Disadvantages**
1) Insufficient availability of suitable breeding stock
2) Tough to apply in shared settings
3) Inadequate technical knowledge, especially documents

## 5.3 Choosing and Eliminating

The process of selecting animals with desirable traits to be the parents of the next generation is called selection.

Culling is the practice of eliminating from the flock any unproductive animals, such as elderly goats, animals with persistent illnesses, and animals with poor mothering skills.

## 5.5 Ratio of Mating

- In a regulated mating system, a male goat should spend 36–42 days running with females. The rationale is that a female who misses her opportunity to mate or go into heat has another shot within the allotted time. 40–50 females might be assigned to a mature buck for service. You may offer a young buck 25–30 females. The physical state of the male and female at mating determines how effective they are.

5.4 Calendar of breeding

A calendar that farmers may use to schedule their flock breeding cycles is provided below. This aids in the farmer's scheduling of input purchases, market trips, and strategic activities..

| Month 1 | 2 | 3 | 4 | 5 | 6 | 7 | 8 | 9 | 10 | 11 | 12 |
|---|---|---|---|---|---|---|---|---|---|---|---|
| Selection of breeding stock | Mating starts (Putting the buck to the females for 42 days) | End of mating | | | Separate the pregnant and the non-pregnant | Supplement and vaccinate against pulpy kidney all pregnant females | Kidding starts | End of kidding | Care of kids | Vaccination against Pulpy kidney Weaning | Flushing of females |
| Flushing | | | | | | | | | | | |
| **Routine management of the flock-Dipping, dosing, vaccinations** ||||||||||||

# 6.0 Kid management

> **At the end of this session farmers should be able to:**
>
> Recognize the benefits and drawbacks of the various kidding seasons.
> - Be familiar with suggested childrearing techniques.
> - Gain knowledge of the significance of weaning;
> - Develop proficiency with various castration techniques

Taking appropriate care of children is crucial for lowering mortality and increasing the growth rate of children. A decrease in child mortality leads to an expansion of the flock, which in turn increases offtake.

## 6.1 Seasons of kidding

• To ensure that the does produce adequate milk for the kids' survival, the moment of birth should match with periods of sufficient feed availability. This often occurs between December and February. Goats may give birth in the winter, when the range is not in ideal condition. In situations like these, it's crucial to make sure the doe is getting enough food and is producing enough milk.

## 6.2 Raising Children

1. Setting Up • Make sure the play space is tidy and has dry bedding (hay or Stover). The advantage of separating pregnant does from the rest of the flock is to ensure an undisturbed birth process and foster good bonding between the doe and kid. The doe may be kept in the kidding area for a few days prior to kidding. Goats that are about to kid exhibit restlessness, separate from the flock, and discharge mucus.

**2. At the time of birth**

The doe must clean and groom her young and be alone for two to four hours in order to foster bonding.

When to step in during the delivery process: • When the mother presents herself incorrectly or has trouble kidding.
• Cut off the membrane covering the nose when the youngster stops breathing or bleating because the doe neglected to clean it.
• Making a navel cut and dousing with iodine. Applying iodine is not required if the bedding is clean.
• When the doe and the young do not form a relationship

## 4. Kid Housing

For the first several weeks to a month or so, keep the youngsters at home (particularly if they must walk considerable distances for water and browsing). For the first four weeks of their lives, the children need warm, dry environments. Children should be shielded from the heat, the cold, and even the spread of illness by their housing.

The Kid boxes are an illustration of kid housing. The child box is 500–600 mm in length, 400–500 mm in width, and 300–400 mm in depth. It is constructed of bamboo or wood. The box's bedding has to be maintained fresh and tidy. This facilitates the detection of diarrhea. After three days, the child may be removed from the box.

## 5. Feeding Kid

• Within the first six hours of life, babies should nurse from the first milk (colostrum-umthubi), which is high in antibodies and boosts the child's immunity. It is advised to bottle feed or foster the doe (ukumunyisela) if she is not generating enough milk for her young.
• At three weeks of age, children begin to nibble on leaves and grass. This is crucial for the growth of the rumen. At the latest one month, you should be able to let them explore and feed. Around 6-7 weeks is when effective grazing and browsing begin.

## 6. Recognition

Individual animal identification is crucial because it facilitates record keeping. There are many approaches that may be used. These include giving animals names, tagging their ears, and notching their ears. When exporting livestock and cattle products, the government additionally mandates that all of the animals have standardized identity for traceability.

### (a) Tagging ears

• The numbering scheme should make sense if numbers are combined with tags; for instance, birth year, sex, and order may all be included. For instance, if a male animal born in 2007 has kid number 23 in the flock, it might have the number 07123, which indicates that the male's birth order is 23, the year of birth is 07, and the male's birth year is 1. Females may display a zero on their tags to indicate their sex.

• Ear tagging is simple and fast. Metal or plastic may be used to make tags. This method's drawbacks include the possibility of the tag coming loose from the ear and the difficulty of quickly reidentifying the animal in a big flock. Avert this issue by attaching tags to both ears.

.

Plastic tags (can come in various shapes, size and colors)

Metal tags

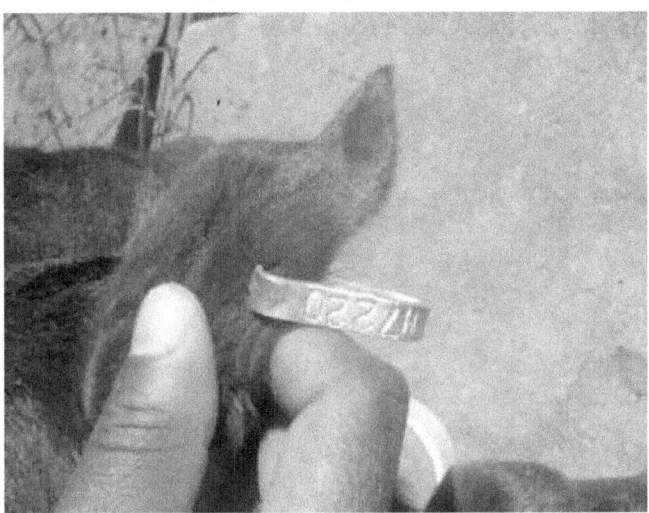

**(b) Ear notching**

• To do this, carve V-shaped indentations on the ear. A V notch's location corresponds to a certain number.
• The fact that it cannot be exclusive to one farmer, particularly in situations where communal grazing occurs, and the farmer would have to hold the cow before reading the notch, is a drawback.
• Units are represented by the left ear and tens by the right ear..

**7. Health care in kids**

Disease incidence will decrease in a clean environment. A farmer should always be aware of respiratory issues, such as coughing or nasal discharge, and diarrhea.

As they say, "Prevention is better than cure."

Ensure that infants get colostrum within six hours of birth; ensure that bedding is clean and dry; avoid confining several children in a compact space; steer clear of moist environments and extreme heat or cold; and refrain from overfeeding infants with milk, since this may lead to scours.

Assure the following for the benefit of the children's overall health:
• Dry areas for sleeping
• Pure drinking water (about five liters per animal daily)
• Sufficient nutrition (3–5% of body weight daily)
• Management of parasites, both internal and external

## 8. Harvesting

• Don't let youngsters wander aimlessly in hazardous areas; instead, make sure they are sheltered to keep them safe from eagles, jackals, and other predators.

Healthy kids

Alternative kid houses

## Weaning

• When the children weigh between 8 and 12 kilos and are, on average, 100 days old, this should be done.
• Complete separation of the does and the youngsters is the most popular weaning technique used with goats. Vaccinating the youngsters and does against pulpy kidney (PK) before to weaning is crucial, however, since this causes stress and increases their susceptibility to PK.
• The process of weaning allows the does to attain optimal physical health in anticipation of the next mating season.

## 9. Castration

To prevent the animal from mating with the females, the spermatic cords are severed or cut. By minimizing the odor of the whole male, castration enhances the quality of the meat. Goats may be castrated using three primary methods: burdizzo, knife/razor, and rubber ring.

### (i) Rubber ring technique

Within the first two weeks of life, the rubber rings are utilized. The ring is stretched and applied over the spermatic cords using an elastrator.
• The child should be held by one person with its rump resting on his knee, its right and left legs in his right and left hands, respectively. That makes it simple to access the scrotum. • One must ensure that the bottom portion of the scrotum is drawn in both testicles.
• Place the rubber ring over the scrotum using an elastrator.
A few weeks later, the scrotum would shrink and fall off. This is a fast and simple approach to apply. Its benefit is that it doesn't need to be disinfected; but, if the rubber ring and testicles fall out, there might be a risk of screw worm infection.

A rubber ring is opened using an elastrator. Put a rubber band around the spermatic cords

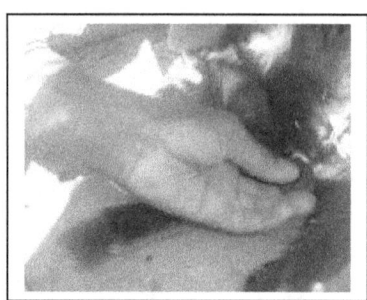

### (ii). Knife/Razor

- Three weeks to three months may pass while doing this. It is necessary to use a sharp knife or razor and to sterilize it in boiling water or an antibacterial solution.

Maintain the animal in a seated position; clean the scrotum with an antiseptic solution; cut open the lower end of the scrotum with a sharp knife or new razor blade; gently remove the testicles from the scrotum and gently rub the upper part of the scrotum to prevent excessive bleeding; cut the spermatic cords; and dip the entire scrotum in an iodine or antiseptic solution before applying wound powder.

## iii. Brudizzo

The spermatic cords are compressed with the burdizzo, causing the testicles to wither over a few days without causing any harm to the scrotum's exterior. The child benefits from this the greatest when they are older than three months.
Clamp the side of the scrotum above the testicle after drawing one testicle down it to crush the spermatic cord. (View the image below.)
- Gently pull each spermatic cord individually.

A burdizzo                    Castrating using a burdizzo

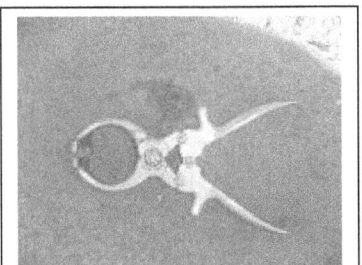 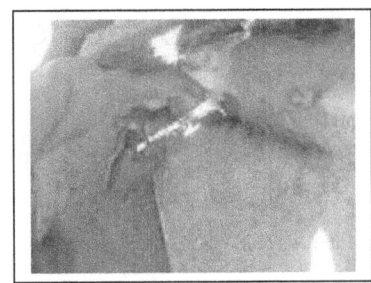

**Other methods**
- Use of a hammer
- Biting with teeth

These techniques cause the animals a great deal of suffering, hence they are not advised. They also raise the possibility of illnesses from animals infecting people.

# 7.0 HUSBANDRY PRACTICES

**Objectives**

> **At the end of this session farmers should be able to:**
> • Recognize the various goat housing systems and their characteristics.
> • Build suitable housing structures for goats;
> • Determine the age of the goats by teeth to inform management choices

## 7.1 Housing

The primary justifications for housing goats:
1. To facilitate management;
2. To reduce childhood and adult fatalities
3. To reduce theft and predation

Goats should be kept inside to protect them from inclement weather, such as rain and sun. Wind. A 1.5 square meter floor area should be allotted to each adult goat. For instance, a home or corral for ten goats should be 1.5 by 10, or 15 square meters.

**Types of housing**

**(a). Walled and Roofed**

**Attributes**

It is normally one meter high, well ventilated, shields animals from the rainy season, is simple to clean, and is warm. In some situations, the wall may be constructed of pole and dagga, and the roof may be built of thatch grass or stover. This may be made to work with varying flock sizes.

**(b) Raised floor with wooden walls, flat roof and a feeding area**

**Attributes**

Warm and simple to clean; the animals may be fed at the pens; the floor is well-drained, minimizing the risk of foot rot. Typically, this works well for small to medium-sized flocks.

**(c) Poles only with no roof**

Features: • Well-ventilated;
• Cheap to build;
• Animals are exposed to heat, cold, and draughts;
• Floors become moist and increase foot rot instances.
It is possible to enhance these buildings.

7.2 Dentition

Goat ages are often ascertained by dental examinations. Goats have eight front teeth, or incisors, in their lower jaw but none in their upper jaw. Goats have huge teeth called molars in the rear of their mouths, which they utilize for chewing.

- The front teeth of animals less than a year old are tiny and pointed. The stage of milk teeth is this.
- After a year, the middle set of teeth fall out and are replaced by two larger ones. The two tooth stage is this one.

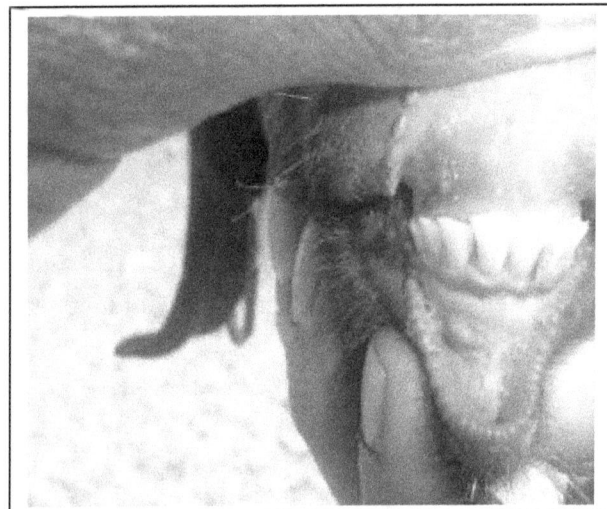

Two tooth stage

- The next two little teeth, one on each side of the initial central pair, fall out and two huge teeth erupt around the age of two years. The fourth tooth stage is this one.

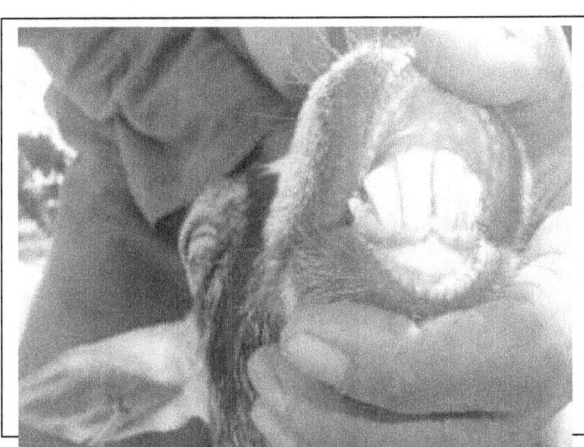

Four tooth stage

- • At the six tooth stage, two huge teeth erupt at the age of three to four years after the following set falls out.

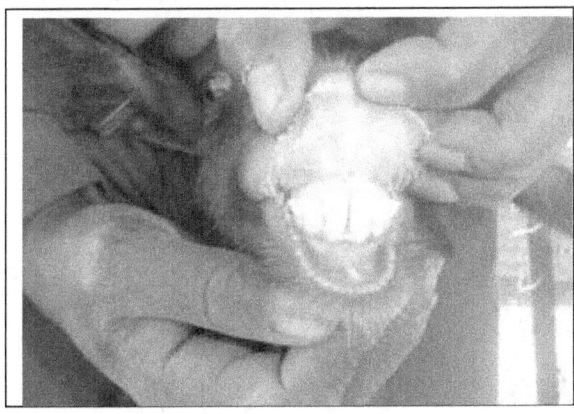

Six tooth stage

- • The goat will have eight permanent teeth by the time its final two milk teeth fall out and two huge ones erupt at the age of four to five. The term "full mouth stage" describes this.

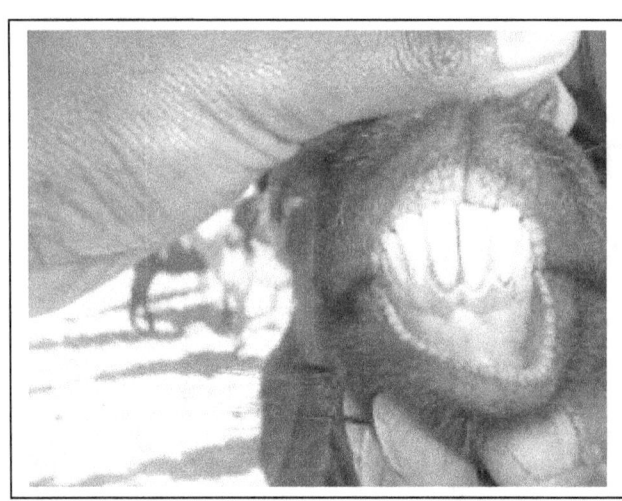

Full mouth

- • The animal's teeth begin to wear down, disintegrate, get loose, and eventually fall out as it gets older (from six years old).

Front teeth worn out at adult stage

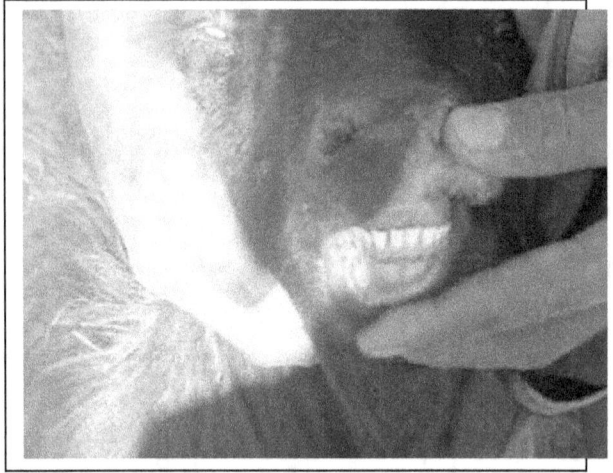

All teeth have dropped at old age.

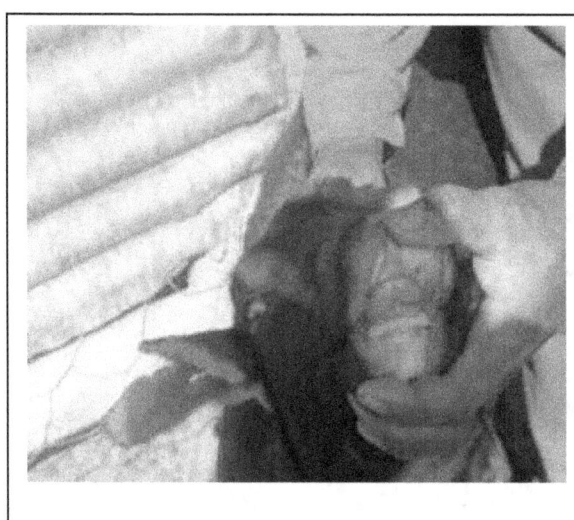

Broken mouth

**Management Tips**

The castrates may be sold at two to six teeth. Culling should begin for bucks and does at the broken mouth stage.

# 8.0 GOAT NUTRITION

**Objectives**

**At the end of this session farmers should be able to:**

- Recognize the sources of the nutrients that goats need. Know how to feed and nourish them. Recognize the digestive system of goats. Determine whether fodder crops are appropriate for semi-arid regions. Grow and preserve fodder crops for use during the dry season.

Goats graze in addition to being natural browsers. They do, however, exhibit selective eating behavior and thrive in environments when a range of foods are provided. Their primary food sources include grass, bushes, and wild berries and pods.

## 8.1 Digestive system

One must be familiar with goats' digestive systems in order to comprehend how they are fed. The graphic below shows that, similar to other ruminants (sheep, cattle), goats have four stomachs: reticulum, omasum, abomasum, and rumen.

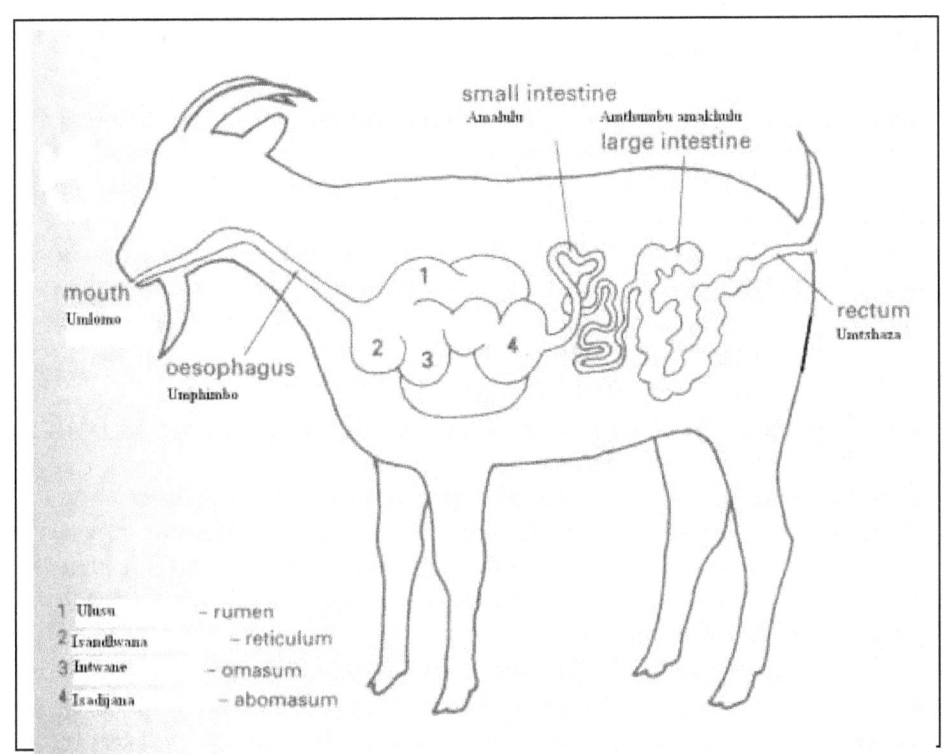

## Feed requirements

A goat's feed intake is determined by its age, breed, sex, size, and physiological state (either breastfeeding or pregnant).
Goats require additional feed during pregnancy and lactation in order to generate milk and support the growth of the fetus. Young goats will eat comparatively more than older goats. Everyday dry matter intake for goats should be between three and five percent of their body weight.

A balanced diet that includes water, carbs, protein, vitamins, minerals, and fiber is necessary for goats. The nutrients and some of the foods from which they may be derived are shown in the table below.

| NUTRIENT | SOURCE |
|---|---|
| Protein | Leguminous plants, Poultry litter, Cotton seed cakes, |
| Carbohydrates | Cereals(maize, sorghum, millet, corn),molasses |
| Vitamins | Vegetables, green forage |
| Minerals | Agro-industrial residue, limestone flour |
| Water | Water bodies, succulents(water melons, cacti, etc) |
| Fibre | Crop residues, hay |

**Types of feeds:**
Blended meals
Unprocessed feeds Supplements

Issues with eating

Swelling

Bloat, or the buildup of gasses in the stomach, is brought on by feeding leguminous foods that are rich in nitrogen. Animals may perish if medical attention is delayed. Acidemia
stones in the bladder poisoning by plants (Umphaphapha)

# 8.2 Production and Preservation of Fodder

The primary obstacle to cattle production is the scarcity of enough fodder, particularly during the dry season. Throughout the year, the rangelands do not provide enough feed, both in terms of quantity and quality, to sustain goat production. As a result, fodder crops must be produced in order to provide additional nourishment during the dry season.

**Fodder crops**

These are crops planted for the purpose of feeding cattle. They may be kept or fed when still fresh. A few instances are included in the table below

## Fodder crops classification

| Class | Crop name | Varieties | Planting | Preservation | Usage |
|---|---|---|---|---|---|
| Grasses | Sorghum | -Sugar drip<br>-Sugar graze | Sow seeds with the first effective rains Spacing-90x20cm | -Harvested at milk dough stage -Make silage. Add legumes to the silage | Refer to the local AREX extension officers. |
| | Millet | -Nutrifeed | -Sow seeds with the first effective rains Spacing-90x20cm | -Harvested at milk dough stage -Make silage. Add legumes to the silage | Refer to the local AREX extension officers |
| | Bana grass | - | -Planted in furrows/rows with the first effective rains<br>-Rows should be 1m x1m in irrigated lands and 1.5mx1m in dry lands -<br>Use plant cuttings (vegetative propagation) | -Allow the plant to grow for one year before it can be harvested -Thereafter harvest when the plants reach 1m and maintain a height of 10-15cm above the ground.<br>-Continue to harvest for the next 3 years<br>- Make hay or silage | Refer to the local AREX extension officers |

| Class | Crop name | Varieties | Planting | Preservation | Usage |
|---|---|---|---|---|---|
| Legumes | Cowpeas, Dolichos bean, Velvet bean | | -Sow seeds with the first effective rains -Spacing-10cmx10cm -For Dolichos the spacing is 75cmx15cm | -harvest after flowering but before hard dough stage before they lose lots of leaves -Mix with cereals for silage making -May harvest them when the seeds have matured. -Crush seeds and mix with cereals. | - Refer to the local AREX extension officers |

| Class | Crop name | Varieties | Planting | Preservation | Usage |
|---|---|---|---|---|---|
| Forage tree | Leucaena | -*Leucacephala* -*Pallicida* | -Scarify the seeds or soften the coat of the seed before planting. -Raise plants in a nursery -Transplant them when they are 20-30cm -Spacing-5mx5m | -Cut, wilt and feed -Cut, dry and feed -Cut, wilt and include in silage mixtures | - Refer to the local AREX extension officers |
| | Acacia | -*Anguistissma*- | -Scarify the seeds or soften the coat of the seed before planting. -Raise plants in a nursery -Transplant them when they are 20-30cm -Spacing-5mx5m | -Cut, wilt and feed -Cut, dry and feed -Cut, wilt and include in silage mixtures | - Refer to the local AREX extension officers |

## Fodder conservation

**Reasons for conserving fodder are:**
- To provide a year-round supply of high-quality feed for animals.
- To keep cattle fertile and producing milk.
- Prevent fatalities and maintain excellent physical health.
- To reduce animal stress by looking for food.

## Preservation techniques

Zimbabwe has two main techniques for conserving fodder: silage and hay production. In the smallholder sector, it is also customary to save agricultural wastes.

## Making Silage

Silage is a product made by carefully fermenting green, succulent crop material in a sealed silo that has a high water and sugar content.
A drum or a pit lined with plastic might be considered a silo.
- A plastic bag.

To ensure that all air is forced out and the fodder within ferments, the silo must be tightly shut and its contents cut and tightly packed.
- Inappropriately fermented fodder rots, is unpleasant, and poisonous; properly ensiled fodder has energy and protein; bacteria transform part of the plant's carbohydrates into appealing tasting lactic acid, which keeps molds and spoilage bacteria from causing the fodder to rot.

## The Method of Plastic Bags

Prior to the start of ensiling, the room has to be inspected annually.
- 15 kilogram plastic bags must to be kept clean since they are often used.
- Roughly chop the dirt-free material to a size of 15-20 mm. • Completely seal the material in the bags to ensure that all of the acid is kept in.

Silage should be available in three weeks if stored in a dry, room-temperature area away from rodents.

Once opened, the whole bag may be fed, lowering the possibility that the remaining fodder would deteriorate. Bags are portable and simple to store. When compared to the pit approach, it also minimizes workload.

**Storage upon preparation**

• Empty bags need to be properly cleaned, dried, and kept in a secure location so they may be used the following year. • Bags of silage should be stored in a room away from ants and mice.

**The Pit Approach**

• To facilitate the water drums' easy entrance and escape, dig a trench that is 2 meters deep, 1.5 meters wide, and 3 meters long. One end of the pit should slope.
• The pit is excavated in an area such as an uphill slope where the water table is not close to the surface.
• The sides of the pit must be perfectly flat with no bumps or rock outcrops, and the bottom wall should slope slightly inward to prevent pockets of air from forming at the edges due to silage settling.
To enable surface drainage and runoff, trenches have to be excavated on both sides of the pit.
• The pit has to be filled as soon as possible and covered with plastic sheeting that is neatly tucked in at the sides. • The chopped length of fodder material should not be longer than 20 cm. • It should be compacted as thoroughly as possible using heavy water drums pushed or rolled over each layer.
• To prevent water seepage into the silage pit and to facilitate drainage, the pit should retain its original form.
• Give it a three-week period to ferment. For mass manufacture, it works well.

## Hay producing

• Summertime abundance of excess grasses and legumes may be stored and used in the winter and during drought spells.
• When they are young and fragile and have enough minerals and vitamins, they should be trimmed throughout the growth season.
• To prevent losses, the dried hay should be placed on a well built hay rack after the grass has been taken out in dry weather, allowed to wilt, and then stacked in little bunches.

# 9.0 Crop- Small Livestock production systems.

**Objectives**

> At the end of this session farmers should be able to:
> - Understand the co-existence between crops and livestock
> - Improve outputs from their crops and livestock enterprises by making use of the relationship between crops and livestock

The interdependence of crops and livestock, which benefits both, is explained by integrated crop-livestock production systems.
- The majority of disadvantaged and small-scale farmers rely mostly on their animals and crops for their income.
- These goods are often thrown away when money is desperately required to pay for other services.

- • Most resource-poor farmers may see an improvement in agricultural production with integrated systems.

## Crop –livestock Interaction

| Benefits of Livestock to Crops | Benefits of Crops to Livestock |
|---|---|
| • Sold to procure crop production inputs | • Crops sold to procure inputs for livestock |
| • Supply of manure for crop production | • Provide feed. |
| • Nitrogen supply through urine. | • Produce Oxygen used by livestock |
| • Livestock helps balance ecosystems through foraging | • Use of crop residues as bedding and roofing material. |
| • Help in seed dispersal of certain crop and grasses | • |
| • Insurance against of crop failure | • |

# 10.0 HEALTH

**Objectives**

**At the end of this session farmers should be able to:**

- Recognize the differences between a sick and a well animal.
- List common illnesses in goats along with prevention and treatment methods.
- Recognize the value of dipping and the many techniques for dipping.
- Recognize the significance of immunization and dosage;
- Recognize the significance of hoof trimming.
- Gain proficiency using syringes and needles to give medication.

- • Diseases lower animal performance and lead to high goat mortality rates.
- Therefore, it's critical that a farmer keep a constant eye on the flock. This makes it easy for the farmer to identify any ill animals and provide care as soon as possible.
- • Early intervention lowers the likelihood that the illness may spread.
- • It's crucial to remember that although certain illnesses may have similar symptoms, they shouldn't be mistaken for one another.
- It is necessary to get a precise and correct diagnosis before trying to treat the animal.
- If you are unsure or notice that your goats are acting strangely, it is advised that you speak with the veterinary officer in your area.
- 
- These easy methods may be used to maintain the flock's health: providing fresh, clean water; sufficient food; and housing that is dry, warm, and well-ventilated.
-

- An unwell animal will often exhibit the following symptoms: dullness of coat; ruffled hair; loss of appetite; drooping ears;
- Dull and pale eyes;
- Difficult in movement; • Dropping tail; and
- Going off feed.

A farmer should have a simple veterinary kit containing the following items

- One bottle (100ml) of antibiotic
- Wound powder (100g)
- Healing oil
- Needles
- Surgical blade
- Iodine (100ml)
- Syringe (20ml)
- Broad spectrum dosing remedy ( 100 ml)
- Cotton wool
- Clinical thermometer
- **Plastic gloves**

**Guide to diagnosis, prevention and treatment of some common diseases**

| Disease /Parasites | Causes | Signs and Symptoms | Treatment | Prevention |
|---|---|---|---|---|
| **Pulpy kidney (Isimeme, umkhuhlane wegazi)** | Caused by a bacterium which is aggravated when there is a sudden change of diet or when the goats are stressed | Unsteady gait and convulsions<br>Animals found dead without showing any signs<br>**At post mortem**<br>Soft pale kidneys<br>Kidneys may look bloody<br>You may see gas filled red intestines (this may also be seen in animals which have been dead for a while)<br>Increased amount of fluid around the heart, which gets thicker and like jelly when sac is opened | When Pulpy Kidney is suspected use antibiotics | Do not change feed suddenly<br>Strategic vaccination is the best way of prevention |

| Disease /Parasites | Causes | Signs and Symptoms | Treatment | Prevention |
|---|---|---|---|---|
| **Heart water** | Caused by blood parasite. The bont tick transmits the parasite. This tick is found mainly in frost-free drier parts of the country, so heart water is mainly found in these areas. | Sick animals may have temperature of 40°C or higher Strange behaviour, for example the goat may turn its head towards its body in a strange manner. Nervous signs such as a high stepping walk, convulsions or kicking very hard. Goats that are very sick with heart water may die.<br><br>**Post mortem**<br>Froth and fluid from the nose. Fluid in the belly, chest and sac surrounding the heart, Swelling of the lungs with froth, and fluid in the windpipe. | When you notice signs of the disease, treat immediately with a broad spectrum acting antibiotic | Dipping to control ticks is recommended. Keep domestic animals away from wild animals |

| | | | | |
|---|---|---|---|---|
| **Coccidiosis (Isihudo)** | Caused by a type of a single celled organism. This disease happens when there are dirty conditions in the animal pens, sleeping areas and kraals. Young animals get this disease very easily. | Watery diarrhoea Dehydration Loss of appetite Loss of condition **Post mortem** There may be spots on the surface of the intestines. When the intestines are cut open, they have a bumpy appearance. | Separate all sick animals Treat all sick animals with a remedy for Coccidiosis Mix ½ teaspoon of salt and 6 teaspoon of sugar in 1 litre of clean warm water. Give the dehydrated kid ¼ to ½ litre of the solution 4 times a day for 3days. | Make sure that you keep the animal pens, sleeping areas and kraals dry, clean and well ventilated. Do not crowd animals into an area that is too small. |

| Disease /Parasites | Causes | Signs and Symptoms | Treatment | Prevention |
|---|---|---|---|---|
| Liver fluke | **How do animals get liver fluke?** The adult fluke lays eggs which hatch in water or wet pasture, giving rise to immature flukes, which cling to the plants growing around marshes and vleis and are swallowed when the goats graze there. | Pale mucous membranes Weight loss. Bottle jaw, which is a soft swelling under the chin of the animal. **Post mortem** Bleeding in the liver. Thickened bile ducts in the liver. Firm, lighter areas in the liver (fibrosis) Liver flukes in the bile ducts. | Use a registered de-wormer in your animals in early spring, in midsummer, and in late autumn or early winter. If fluke infection is serious; animals may need additional treatments during summer. If you have been treating for liver fluke and there is no improvement, then you need to ask your veterinarian or animal health technician for help. | Where possible, fence off vleis streams and dams to stop the goats going there. Fence off the pastures that are known to give liver fluke problems. They should be grazed only in the winter months, when the fluke numbers are much lower. Strategic dosing |

| | | | | |
|---|---|---|---|---|
| **Roundworms (izilo zesisu)** | Goats get roundworms when they take in the immature worms while eating grass. These immature worms grow into adult worms in the animal. Young animals are most badly affected | You may see bottle jaw. The inside of the eyelids could be pale. Diarrhoea may occur but remember diarrhoea may also have other causes (such as Coccidiosis or toxic plants). During winter or the dry season, animals may be in poor body condition. **Post mortem** There may be bleeding or having worms on the stomach or intestinal lining. | If you see the signs treat with a worm remedy. | Have a flexible dosing programme |

| Pneumonia (Isihlabo) | Caused by a bacteria<br>Usually occurs if goats are under stress due to exposure e.g. to wind, cold and heavy rain<br>Animals usually develop the disease after travelling for long distances | Animals may seem tired and walk behind the rest of the flock May stop eating properly<br>High temperature<br>Animals show fast breathing and breathe with difficulty<br>Mucus discharge from the nose<br><br>**Post mortem**<br>The lungs look patchy with red patches and normal pink areas<br>Large part of the lung will be firm and red in colour<br>Lungs may be covered with white layer which sticks to the inside of the ribs<br>Froth in the windpipe<br>If put in water the lungs will sink instead of floating | Treat with a long acting antibiotic product | Provide shelter all the time During long journeys allow goats stops to rest, eat and drink |
|---|---|---|---|---|
| **Disease /Parasites** | **Causes** | **Signs and Symptoms** | **Treatment** | **Prevention** |

| Orf (izilonda emlonyeni) | Caused by a virus found in the soil.

This virus gets into the animal through a cut in the skin

An infected kid can spread the disease to its mother during suckling | Small round scabs seen usually at the corner of the mouth

These scabs spread to the muzzle, nose and eyes

Encrusted sores may develop on the teats of suckling females | The disease usually clears on its own
Apply petroleum jelly to keep the scabs soft
Cannot be treated but you can spray with an aerosol antibiotic to avoid
secondary infection
**NB** always wear gloves as this can be transmitted to humans | When a few animals are affected, vaccinate the healthy animals

Do not vaccinate healthy animals when there is no orf in the flock

Kids should be bottle fed when affected to avoid spreading the disease to its mother |
|---|---|---|---|---|

| **Abscesses** (Amathumba) | Caused by bacteria found in the dust or manure. Usually develops from injury caused by ticks, thorns or wire | Round swelling which maybe red and painful on touching. Usually develops in front of the shoulder on the head or neck or on the flank on the hind quarter, but can also develop on other areas on the body. | Should be done after hair has fallen off and there is a soft spot in the middle. For hairy goats, shave and cut a cross over the soft spot. Use your finger to squeeze out puss. Clean the wound with boiled salty water. Use a suitable wound spray to keep away flies (If this is not possible use some herbs that repel flies). If possible give an antibiotic injection | If the animal has several bad abscesses or often gets abscesses it should be culled. Control ticks |
|---|---|---|---|---|

For other diseases consult your local veterinary office

bghhghcf

## ROUTINE HEATH MANAGEMENT PRACTICES

**Regular health management procedures include of immunization, dosing, dipping, and clipping of hooves.**

### Dipping

External parasites like mange mites and ticks are responsible for a multitude of ailments. Using acaricides to control these parasites is the most efficient strategy to avoid these disorders. For goats, there are many techniques for applying dips.

### Pour On
- The acaricides are applied to the animal's back using tiny containers; the amount used depends on the animal's weight.
- As a result, the acaricide disperses throughout the animal when sweating, eliminating all exterior parasites from its body.
- This procedure becomes exceedingly difficult in big flocks since it needs handling each animal individually. It is advised during the dry season and when a small number of animals are afflicted by ticks.

### Greasing

- The most popular kind of acaricide used in this situation is tick grease, which is administered directly to ticks, commonly beneath the tail, on the udder, and on the ears. This is also often used in cases when certain animals have ticks.
- Tick repellents may be made from certain greases.

### Spraying

- The animals sometimes go through a spray race in which they are sprayed all over their bodies. The spray that emerges from the nozzles will contain the acaricide.
- The main issue with this dipping technique is that sometimes the nozzles clog, leaving the animals without a enough amount of spray.
- The knapsack is sometimes used to spray the animals.

### (iv). Plunge dip

- The animals swim through an acaricide-filled plunge dip in this method of dipping; the acaricide comes into touch with the animal's whole body, killing all parasites on the body.
- Since it doesn't involve touching the animals and isn't labor-intensive, this is advised in big flocks.
- In the event that the dip tank is not yet built, the goats may be plunged using half-drums. However, it is imperative that the dip solution be disposed of properly to protect the environment.
- Small ruminants often use dip tanks with a 4266 liter capacity.
- The dip tank ought to empty smoothly.

How often do you dip?

- Because of the high tick load in the summer, dive once a week; in the winter, dip once every two weeks; nevertheless, try to dip during the warmer hours of the day to prevent pneumonia.

A common tiny ruminant's plunge dip tank.

- Goat farmers should be able to bargain for costs that are commensurate with the quality of the animal.
- Farmers can come together as a group to improve their buying power.
- Farmers should gather up to date information about market trends.
- Produce good quality goats in the right amount (optimum production).
- .Farmers should avoid desperate/ grief selling.
- The farmer can sell straight or sell at a sale.

**Financing the Goat Business**

Most farmers lack understanding of how much they need for their goat businesses. Farmers should have an idea of how much they require for start up costs and running fees.

They must prepare a cash plan/budget. This will help the farmer to look for funds. The plan should state how much money is needed for the following items:
- Infrastructure • Breeding stock • Feeds • labour • Veterinary supplies • Transport

It should also estimate income from the business.

Sources of funds are: • Own savings: which is usually cheap but not easy to raise.

- Loans from business banks: These are very expensive and not readily available to most country goat farmers. The requirements for these loans are usually strict and hard. The requirements include among other things: • Track record • Formal establishment of company or firm. • Financial details • Collateral.
-

Institutions that provide short term facilities include commercial banks like Agribank, ZABG, Premier Banking Corporation, Commercial Bank of Zimbabwe, development institutions like the SEDCO and the Infrastructural Development Bank and a range of lower level financial institutions such as microfinance institutions and village banks or savings and credit cooperatives. Issues funded through short term facilities are of a working capital type such as food, medical medicine, breeding stock etc.

• Group lending: A plan whereby groups containing approximately five to fifteen smallholder farmers or rural businesses come together to borrow money from the bank. These should be staying within the same area. They are bound by a group contract and run a group bank account. They should have similar project goals for them to qualify for the loan. The group will have joint responsibility on the group loan given

• Credit plans: There are ancient schemes where groups loan each other animals. These are rare. There are plans that are government driven on farming products. However most of them focus on food farming.

• Donors: offered only for poor farmers for restocking activities. These are cheap funds. They are open for group projects. However, these funds are not usually enough to run successful businesses.

• Contract farming or out farmer schemes: are relationships in which buyers of agricultural goods give funds (either in-kind or in cash) to producers. The loan is usually tied to a buying deal. This plan is not yet offered in the goat sub sector. The out grower plan is active in the cattle, pork, poultry and ostrich sub sectors. The companies provide farmers with inputs and take the equal amount plus interest from the farmer on providing the products. Contract farming and out farmer plans help producers to gain access to high-value markets, as well as to improve their output by giving them loans with integrated services such as technical and marketing support.

## Dosing/Drenching

- The goat is being forced to swallow liquid medication, which is often done to manage intestinal parasites.
- • When dealing with big flocks, a graded syringe and a dosing gun attached to a two-liter container are often used. Using a little syringe, the medication may be extracted from the container for small flocks.
- • A long-necked bottle is sometimes used for drenching. It's crucial to use care while giving your pet a soak.
- • The animal should be given the syringe, gun, or bottle so that the liquid flows into its mouth gradually and is ingested.

### Hoof trimming

- Animals' hooves get overgrown and need regular clipping to avoid harm when they walk on hard, uneven terrain.
- Either a pair of foot shears or a sharp, curved knife are employed.
- Trim the portion of the hoof that is overgrown. Cut the heels too if they're overgrown.
- Take care not to slice through the hoof too much and expose the living tissue.
- To make the hooves firm and stop them from breaking and rotting, dip them in a solution of copper sulphate. This may be carried out year before to the start of the rainy season.

### Injections

Both vaccination and treatment of some disorders include injections. Three routes are available for injections:

#### (i). Intravenous

This kind of injection is administered via a vein straight into the animal's bloodstream. Usually, this is done to get a prompt reaction and cure certain conditions. This is often done by a veterinary professional. This kind of injection requires the use of a lengthy needle.

#### (ii). Intramuscular

These injections are administered deeply into the shoulder or back leg muscles. Usually, illnesses are treated with this, and a large needle is employed.

#### (iii). Subcutaneous

This kind is commonly administered under the skin in the neck or behind the shoulder. The injection is placed behind a skin fold that has been pulled up. This employs a short needle and is often used for vaccinations.

NB: Syringes and needles should be boiled in water for twenty minutes to sanitize them.

Subcutaneous method of injecting goats

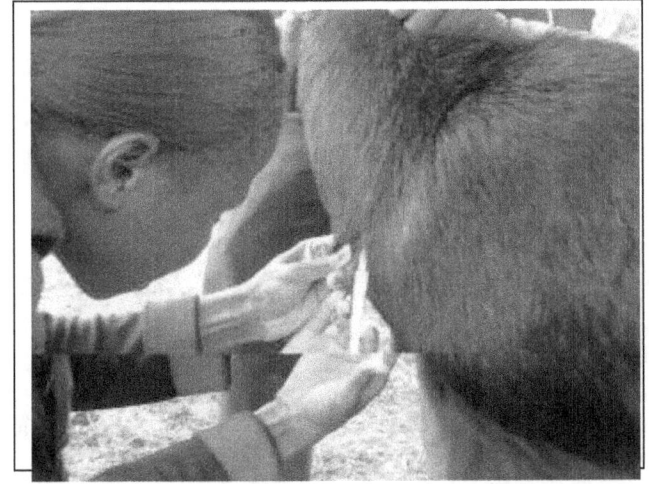

## 11.0 GOAT BUSINESS MANAGEMENT
**Objectives:**

> **By the end of the session farmers should be able to:**
> - Learn about the process of selling goats.
> - Learn about the various market systems and marketing choices that are out there. Know order to sell goats effectively.
> - Know what you need to do to negotiate and have power in negotiations.
> - Know how to make business plans.
> - Learn how to keep records.
> - Learn how to make a budget and get money for your business.

## GOAT MARKETING

### Introduction

When goat farming is done as a business, the main goal is to make money. If you make a good product that meets the needs of the market, you can make money. To put it simply, marketing means figuring out what the buyers want and then giving them a product (goats) that meets those needs in the right amount at the right time and place.

### 11.1 Understanding goat marketing

♣ Identifying needs: Buyers require goats of different ages, size, breeds, etc. Some buyers such as the local sellers are much worried about the size while some buyers from the cities, high value markets stress on quality.

♣ Specific group of customers: Some of the specific goat markets are individual sellers, abattoirs, NGOs, ethnic groups and export market.

♣ Product: In the goat business the goods that we can sell to the market are live goats, goat meat, skins, milk, fur and manure.

♣ Right quantities: It is also important for farmers to be able to plan their production so that they regularly give the needed quantities at stated time intervals (e.g. 250 slaughter goats every month). This is key in business as this helps towards building lasting and mutually useful (win-win) relationships with your buyers.
♣ Right time and place: When we start our goats to organized high value markets, we need to plan our production and logistics to meet the market requirements.

## 11.2 Goat shops

♣ Currently, the market for goats is highly unorganized and brokers rule deals.
♣ Sales are mainly at farm gate level.
♣ There is lack of market knowledge.

There following are some of the market choices open to goat farmers.

### Individual traders

• This market comprises individual buyers who buy goats for sale in high value urban markets.

Private sector companies (Abattoirs, butcheries and others

• These usually require huge amounts of goats to serve both the local market and the foreign market. .
• This market focuses on quality, stability and prompt supply.
• There is high travel costs involved when reaching this market.

**Ethnic groups:**

• The Moslem community offers a market during their religious events. • The goats are killed according to Halaal custom. • There are special butcheries and abattoirs that serve the Moslem community.

Non-Governmental Organizations:

• NGOs working in restocking projects provide a market for breeding stock. • They pay competitive prices.

Export market: • The export demand is found in Asian countries and other African countries. • The market is more demanding in terms of needs.

Why are the buyers giving low prices for your goats?

………………………………………………………………………………

………………………………………………………………………………

What do you think should be done to solve the problems?

………………………………………………………………………………

………………………………………………………………………………

**Marketing Tips**

www.ingramcontent.com/pod-product-compliance
Lightning Source LLC
Chambersburg PA
CBHW062122220526
45471CB00010B/3837